ANIMAL TRACKS
of
ALASKA

Ian Sheldon & Tamara Hartson

LONE PINE

THE PUBLISHER: LONE PINE PUBLISHING

1808 B Street NW, Suite 140	10145-81 Avenue
Auburn, WA 98001	Edmonton, AB T6E 1W9
USA	Canada

Lone Pine Publishing website: http://www.lonepinepublishing.com

Canadian Cataloguing in Publication Data

Sheldon, Ian, (date)
  Animal tracks of Alaska

  Includes bibliographical references and index.
  ISBN-13: 978-1-55105-244-1
  ISBN-10: 1-55105-244-X

  1. Animal tracks—Alaska—Identification.
I. Hartson, Tamara, (date) II. Title.
QL768.S52 1999   591.47'9   C99-911000-4

Editorial Director: Nancy Foulds
Editor: Volker Bodegom
Production Manager: Jody Reekie
Design, layout and production: Volker Bodegom, Monica Triska
Cartography: Volker Bodegom
Animal illustrations: Gary Ross, Horst Krause, Ian Sheldon
Track illustrations: Ian Sheldon
Cover illustration: Grizzly Bear by Gary Ross
Scanning: Elite Lithographers Ltd.

We acknowledge the financial support of the Government of Canada through the Book Publishing Industry Development Program (BPIDP) for our publishing activities.

PC: P1   Canadä

CONTENTS

INTRODUCTION

If you have ever spent time with an experienced tracker, or perhaps a veteran hunter, then you know just how much there is to learn about the subject of tracking and just how exciting the challenge of tracking animals can be. Maybe you think that tracking is no fun, because all you get to see is the animal's prints. What about the animal itself—is that not much more exciting? Well, for most of us who don't spend a great deal of time in the beautiful wilderness of Alaska, the chances of seeing the majestic Moose or the fun-loving River Otter are slim. The closest we may ever get to some animals will be through their tracks, and they can inspire a very intimate experience. Remember, you are following in the footsteps of the unseen—animals that are in pursuit of prey, or perhaps being pursued as prey.

This book offers an introduction to the complex world of tracking animals. Sometimes tracking is easy. At other times it is an incredible challenge that leaves you wondering just what animal made those unusual tracks. Take this book into the field with you, and it can provide some help with the first steps to identification. Animals tracks and trails are this book's focus; you will learn to recognize subtle differences for both. There are, of course, many additional signs to consider, such as scat and food caches, all of which help you to understand the animal that you are tracking.

Remember, it takes many years to become an expert tracker. Tracking is one of those skills that grows with you as you acquire new knowledge in new situations. Most importantly, you will have an intimate experience with nature. You will learn the secrets of the seldom seen. The more you discover, the more you will want to know and, by developing a good understanding of tracking, you will gain an excellent appreciation of the intricacies and delights of our marvelous natural world.

How to Use This Book

Most importantly, take this book into the field with you! Relying on your memory is not an adequate way to identify tracks. Track identification has to be done in the field, or with detailed sketches and notes that you can take

Arctic Fox

home. Much of the process of identification involves circumstantial evidence, so you will have much more success when standing beside the track.

This book is laid out so as to be easy to use. There is a quick reference appendix to the tracks of all the animals illustrated in the book beginning on p. 124. This appendix is a fast way to familiarize yourself with certain tracks and the content of the book, and it guides you to the more informative descriptions of each animal and its track.

Each animal's description is illustrated with the appropriate footprints and the trail patterns that it usually leaves. Although these illustrations are not exhaustive, they do show the tracks or groups of prints that you will most likely see. You will find a list of dimensions for the tracks, giving the general range, but there will always be extremes, just as there are with people who have unusually small or large feet. Under the category 'Size' (of animal), the 'greater-than' sign (>) is used when the size difference between the sexes is pronounced.

If you think that you may have identified a track, check the 'Similar Species' section for that animal. This section is designed to help you confirm your conclusions by pointing out other animals that leave similar tracks and showing you ways to distinguish among them.

As you read this book, you will notice an abundance of words such as 'often,' 'mostly' and 'usually.' Unfortunately, tracking will never be an exact science; we cannot expect animals to conform to our expectations, so be prepared for the unpredictable.

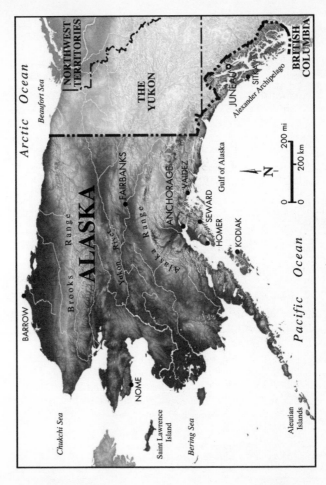

Arctic Ocean

Beaufort Sea

NORTHWEST TERRITORIES

THE YUKON

BRITISH COLUMBIA

JUNEAU ○ SITKA

Alexander Archipelago

FAIRBANKS ●

ANCHORAGE ●

Chukchi Sea

Brooks Range

ALASKA

Yukon River

Alaska Range

VALDEZ ●
SEWARD ●
HOMER ● KODIAK ●

Gulf of Alaska

BARROW ●

NOME ●

Saint Lawrence Island

Bering Sea

Aleutian Islands

Pacific Ocean

200 mi
200 km

—N—

7

Tips on Tracking

As you flip through this guide, you will notice clear, well-formed prints. Do not be deceived! It is a rare track that will ever show so clearly. For a good, clear print, the perfect conditions are slightly wet, shallow snow that isn't melting, or slightly soft mud that isn't actually wet. These conditions can be rare—most often you will be dealing with incomplete or faint prints, where you cannot really be sure of the number of toes.

Should you find yourself looking at a clear print, then the job of identification is much easier. There are a number of key features to look for: measure the length and width of the print, count the number of toes, check for claw marks and note how far away they are from the body of the print, and look for a heel. Keep in mind more subtle features, such as the spacing between the toes, whether or not the toes are parallel, and whether fur on the sole of the foot has made the print less clear.

When you are faced with the challenge of identifying an unclear print—or even if you think that you have made a successful identification from one print alone—look beyond the single footprint and search out others. Do not rely on the dimensions of one print alone, but collect measurements from several prints to get an average impression. Even the prints within one trail can show a lot of variation.

Try to determine which is the fore print and which is the hind, and remember that many animals are built very differently from humans, having larger forefeet than hind feet. Sometimes the prints will overlap, or they can be directly on top of one another in a direct register. For some animals, the fore and hind prints are pretty much the same.

Check out the pattern that the tracks make together in the trail, and follow the trail for as many paces as is necessary for you to become familiar with the pattern. Patterns are very important, and can be the distinguishing feature between different animals with otherwise similar tracks.

Meadow Jumping Mouse

Follow the trail for some distance, because it may give you some vital clues. For example, the trail may lead you to a tree, indicating that the animal is a climber—or it may lead down into a burrow. This part of tracking can be the most rewarding, because you are following the life of the animal as it hunts, runs, walks, jumps, feeds or tries to escape a predator.

Take into consideration the habitat. Sometimes very similar species can be distinguished by their habitats only—one might be found on the riverbank, whereas another might be encountered just in the dense forest.

Think about your geographical location, too, because some animals have a limited range. This consideration can rule out some species and help you with your identification.

Remember that every animal will at some point leave a print or trail that looks just like the print or trail of a completely different animal!

Finally, keep in mind that if you track quietly, you might catch up with the maker of the prints.

Terms and Measurements

Some of the terms used in tracking can be rather confusing, and they often depend on personal interpretation. For example, what comes to your mind if you see the word 'hopping'? Perhaps you see a person hopping about on one leg—

Herring Gull

or perhaps you see a rabbit hopping through the countryside. Clearly, one person's perception of motion can be very different from another's. Some useful terms are explained below, to clarify what is meant in this book and, where appropriate, how the measurements given fit in with each term.

The following terms are sometimes used loosely and interchangeably; for example, a rabbit might be described as 'a hopper' and a squirrel as 'a bounder,' yet both leave the same pattern of prints in the same sequence.

Bounding: A gait of four-legged animals in which the two hind feet land simultaneously, usually registering in front of the forefeet. Common in rodents and the rabbit family. 'Hopping' or 'jumping' can often be substituted.

hind prints   *fore prints*

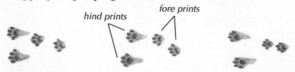

Gait: An animal's gait describes how it is moving at some point in time, and results in observable trail characteristics.

Galloping: A gait used by animals with four even-length legs, such as dogs, moving at high speed, hind feet registering in front of forefeet.

hind prints *fore prints* gallop group

Hopping: Similar to bounding. With four-legged animals, usually indicated by tight clusters of prints, fore prints set between and behind the hind prints. A bird hopping on two feet creates a series of paired tracks along its trail.

Loping: Like galloping, but slower, with each foot falling independently and leaving a trail pattern that consists of groups of tracks in the sequence fore-hind-fore-hind, usually roughly in a line.

Mustelids (weasel family) often use *2x2 loping*, in which the hind feet register directly on the fore prints. The resulting pattern has angled, paired tracks.

11

Running: Like galloping, but applied generally to animals moving at high speed. Also used for two-legged animals.

Trotting: Faster than walking, slower than running. The diagonally opposite limbs move simultaneously; that is, the right forefoot with the left hind, then the left forefoot with the right hind. This gait is the natural one for canids (dog family), short-tailed shrews and voles.

hind print   fore print

Canids may use ***side-trotting***, a fast trotting in which the hind end of the animal shifts to one side. The resulting track pattern has paired tracks, with all the fore prints on one side, and all the hind prints on the other.

Walking: A slow gait in which each foot moves independently of the others, resulting in an alternating track pattern. This gait is common for felines (cat family) and deer, as well as wide-bodied animals, such as bears and porcupines. The term is also used for two-legged animals.

Caribou

Other Tracking Terms:

Dewclaws: Two small, toe-like structures set above and behind the main foot of most hoofed animals.

dew claw marks

dragline

Direct Register: The hind foot falls directly on the fore print.

double register

direct register

Double Register: The hind foot overlaps the fore print only slightly or falls beside it, so that both prints can be seen at least in part.

Dragline: A line left in snow or mud by a foot or the tail dragging over the surface.

Gallop Group: A track pattern of four prints made at a gallop, usually with hind feet registering in front of forefeet.

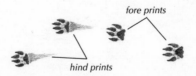

fore prints

hind prints

Height: Taken at the animal's shoulder.

Length: The animal's body length from head to rump, not including the tail, unless otherwise indicated.

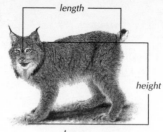

Lynx

Metacarpal Pad: A small pad near the palm pad or between the palm pad and heel on the forefeet of bears and members of the weasel family.

Print: (also called '***track***.') Fore and hind prints are treated individually. Print dimensions given are 'length' (including claws—maximum values may represent occasional heel register for some animals)—and 'width.' A group of prints made by each of the animal's feet makes up a track pattern.

Register: To leave a mark—said about a foot, claw or other part of an animal's body.

Retractable: Describes claws that can be pulled in to keep them sharp, as with the cat family; these claws do not register in the prints. Foxes have semi-retractable claws.

Sitzmark: The mark left on the ground by an animal falling or jumping from a tree.

Straddle: The total width of the trail, all prints considered.

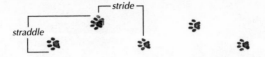

Stride: For consistency among different animals, the stride is taken as the distance from the center of one print (or print group) to the center of the next one. Some books may use the term 'pace.'

Track: Same as '***print***.'

Track Pattern: The pattern left after each foot registers once; a set of prints, such as a gallop group.

Trail: A series of track patterns; think of it as the path of the animal.

Hoary Marmot

MAMMALS

River Otter

Muskox

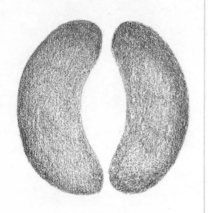

Fore and Hind Prints
Length: 4–5 in (10–13 cm)
Width: 4–5 in (10–13 cm)
Straddle
10–15 in (25–38 cm)
Stride
Walking: 15–25 in (38–65 cm)
Size (males>females)
Height: 3–5 ft (90–150 cm)
Length: 6.3–8 ft (1.9–2.4 m)
Weight
370–900 lb (170–410 kg)

walking

MUSKOX
Ovibos moschatus

   With its stout profile and shaggy coat, the Muskox seems to be heavily burdened, but it is surprisingly agile and can run very quickly when necessary. Herds of 15 to 100 animals can be seen roaming the tundra, grassy river valleys, lakeshores and shrubby hillsides of northern Alaska.

   Each print is a pair of facing crescents, giving an overall circular appearance. The usual alternating walking track pattern shows the hind print on top of or near the easily distinguished larger, deeper fore print, but the front and rear rims of a print are so similar that direction of travel may be difficult to determine. The distinctive, powerful odor of the Muskox may be more obvious than its nondescript prints.

Similar Species: Caribou (p. 22) prints are very similar. Bison (p. 20) prints are similar, but no wild Bison occur where Muskox do. Muskox prints on a hard surface may resemble Horse (p. 30) prints.

Bison

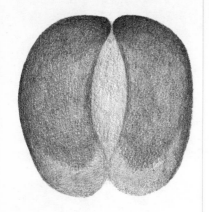

Fore and Hind Prints
Length: 4–6 in (10–15 cm)
Width: 4–6 in (10–15 cm)
Straddle
10–21 in (25–55 cm)
Stride
Walking: 14–32 in (35–80 cm)
Size (bull>cow)
Height: 5–6 ft (1.5–1.8 m)
Length: 10–12 ft (3–3.7 m)
Weight
Male: 800–2000 lb (360–900 kg)
Female: 700–1100 lb (320–500 kg)

walking

BISON (Buffalo)
Bison bison (formerly *Bos bison*)

An estimated 70 million Bison once roamed North America. As few as 1500 survived the wholesale Bison slaughter of the nineteenth century. Today, as a result of a major effort to save this magnificent beast from extinction, up to 100,000 Bison inhabit scattered protected areas and ranches across the continent.

In a Bison's alternating walking pattern, the slightly smaller hind foot usually registers on or near the fore print. On firm ground, only the outer edge of the hoof may register, but in soft mud or snow the whole foot registers—perhaps the dewclaws too. Foot drag is common. The abundant 'pies' may be mistaken for those of domestic cattle. Additional signs of Bison include rubbing posts or trees that have tufts of distinctive brown hair hanging from them, and the large pits in which Bison wallow. Do not be fooled by a Bison's calm exterior: a hefty bull Bison can inflict serious injury!

Similar Species: Domestic Cattle (*Bos* spp.) prints are similar. On firm surfaces, Horse (p. 30) prints can resemble Bison prints.

Caribou

Fore and Hind Prints
Length: 3–4.7 in (7.5–12 cm)
Length with dewclaws:
    to 8 in (20 cm)
Width: 4–6 in (10–15 cm)
Straddle
9–14 in (23–35 cm)
Stride
Walking: 16–32 in (40–80 cm)
Running: to 5 ft (1.5 m)
Group length: to 9 ft (2.7 m)
Size (buck>doe)
Height: 3.5–4 ft (1.1–1.2 m)
Length: 6.5–8.5 ft (2–2.6 m)
Weight
150–600 lb (70–270 kg)

*walking
(in snow)*

*walking
(on hard
ground)*

CARIBOU
Rangifer tarandus

There are few places in North America where you might see the elegant antlers of the male Caribou, a true wilderness animal. The Caribou's sensitivity to human encroachment has reduced its range to remote northern regions and mountain parks. It feeds in groups, above the tree line.

In winter, the soft inner sole of a Caribou's hoof hardens and shrinks, leaving a firm outer wall that makes neat circles on firmer surfaces. The snowshoe-like hooves spread wide, leaving distinctive large, rounded prints. Big dewclaws help the Caribou to distribute its weight on snow. These dewclaws register behind the forefeet, but rarely the hind feet; the faster a Caribou runs, the more perpendicular its dewclaws are to the direction of travel. In snow, foot drag is common—look at the print to see how Caribou swing their legs as they walk. Scrape marks indicate where Caribou have dug for lichen hidden beneath the snow.

Similar Species: The Caribou's print shape and size and its range minimize confusion. In extreme northern Alaska, Muskox (p. 18) prints are comparable, but lack dewclaws. Juvenile Moose (p. 24) prints may be similar.

Moose

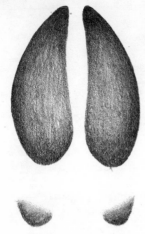

Fore and Hind Prints
Length: 4–7 in (10–18 cm)
Length with dewclaws: to 11 in (28 cm)
Width: 3.5–6 in (9–15 cm)
Straddle
8.5–20 in (22–50 cm)
Stride
Walking: 1.5–3 ft (45–90 cm)
Trotting: to 4 ft (1.2 m)
Size (bull>cow)
Height: 5–6.5 ft (1.5–2 m)
Length: 7–8.5 ft (2.1–2.6 m)
Weight
600–1100 lb (270–500 kg)

walking

MOOSE
Alces alces

The impressive male Moose, largest of the deer, has a massive rack of antlers. Moose are usually solitary, though you may see a cow with her calf. Despite its placid appearance, a moose may charge humans if approached.

The ungainly shaped Moose moves gracefully, leaving a neat alternating walking pattern. The hind feet direct or double register on the fore prints. Long legs allow for easy movement in snow; where the print is deeper than 1.2 inches (3 cm), the dewclaws—which give extra support for the animal's great weight—register but far behind the hoof. In summer, look for tracks in mud beside ponds and other wet areas, where Moose especially like to feed; they are excellent swimmers. In winter, Moose feed in willow flats and coniferous forests, leaving a distinct browseline (highline). Ripped stems and scraped bark, 6 feet (1.8 m) or more above the ground, are additional signs of Moose.

Similar Species: Caribou (p. 22) tracks—much rounder and smaller, with a narrower straddle—may be mistaken for a juvenile Moose's.

Mountain Goat

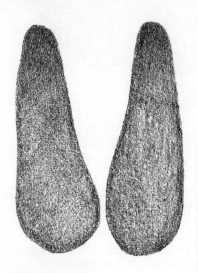

Fore and Hind Prints
Length: 2.5–3.5 in (6.5–9 cm)
Width: 2–3.3 in (5–8.5 cm)

Straddle
6.5–12 in (17–30 cm)

Stride
Walking: 10–19 in (25–48 cm)

Size (billy>nanny)
Height: 3–3.5 ft (90–110 cm)
Length: 5–6 ft (1.5–1.8 m)

Weight
100–300 lb (45–140 kg)

walking

MOUNTAIN GOAT
Oreamnos americanus

Spotting the dazzling white coat of a Mountain Goat is truly a high-mountain wilderness experience. This goat has a preference for high terrain, usually rocky slopes above the tree line, but in severe weather it may descend. The Mountain Goat's keen vision and its remote setting make it difficult to approach; its tracks in snow may be your best clue that it is around.

The goat's long, widely spreading toes produce a squarish print. The hard rim and soft middle of the foot help this agile goat clamber over the most unlikely crags at remarkable speeds, but even the Mountain Goat can make a fatal mistake! In deeper snow, the feet may leave draglines and the dewclaws may register. The alternating walking pattern shows a double register, hind print on top of fore.

Similar Species: Dall's Sheep (p. 28) prints are smaller, narrower and more pointed.

Dall's Sheep

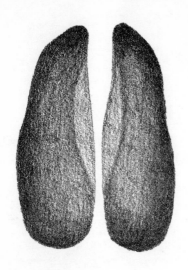

Fore and Hind Prints
Length: 2.5–3.5 in (6.5–9 cm)
Width: 1.8–2.5 in (4.5–6.5 cm)

Straddle
6–12 in (15–30 cm)

Stride
Walking: 14–24 in (35–60 cm)

Size (ram>ewe)
Height: 2.5–3.5 ft (75–110 cm)
Length: 4–6.5 ft (1.2–2 m)

Weight
75–270 lb (34–120 kg)

walking

DALL'S SHEEP (Thinhorn Sheep, Stone Sheep, Kenai Sheep)
Ovis dalli

The Dall's Sheep is found in high, mountainous areas of Alaska. This gregarious animal inhabits areas of rocky terrain, making tracking difficult. In winter it feeds mainly on woody plants that grow between the rocks. Summer is the best time to find tracks, when the sheep move into areas of grasses, forbs and sedges for feeding.

Prints are squarish, but pointed toward the front. The outer edge of the hoof is hard and the inner part is soft, giving the sheep a good grip on tricky terrain. The neat alternating walking pattern is a direct or double register of the hind print over the fore print. When a sheep runs, the toes spread wide. The trails may lead you to sheep beds, which are hollows dug into the snow. They are used many times and often have a large accumulation of droppings.

Similar Species: Tracks of the Bighorn Sheep (*Ovis canadensis*), which is not found in the state, are indistinguishable. Mountain Goats (p. 26) have prints that are wider at the toe and they rarely run.

Horse

**Fore Print
(hind print is slightly smaller)**
Length: 4.5–6 in (11–15 cm)
Width: 4.5–5.5 in (11–14 cm)
Stride
Walking: 17–28 in (43–70 cm)
Size
Height: to 6 ft (1.8 m)
Weight
to 1500 lb (680 kg)

walking

HORSE
Equus caballus

Back-country use of the popular Horse means that you can expect its tracks to show up almost anywhere.

Unlike any other animal in this book, the Horse has only one huge toe. This toe leaves an oval print with a distinctive 'frog' (V-shaped mark) at its base. When the Horse is shod, the horseshoe shows up clearly as a firm wall at the outside of the print. Not all horses are shod, so do not expect to see this outer wall on every horse print. A typical, unhurried horse trail is an alternating walking pattern, with the hind prints registering on or behind the slightly larger fore prints. Horses are capable of a range of speeds—up to a full gallop—but most recreational horseback riders take a more leisurely outlook on life, preferring to walk their horses and soak up the mountain views!

Similar Species: Mules (rarely shod) have smaller tracks. Muskox (p. 18) prints on a hard surface may resemble Horse prints.

Grizzly Bear

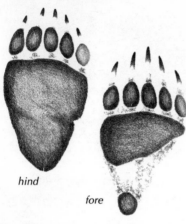

hind

fore

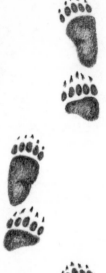

Fore Print
Length: 5–7 in (13–18 cm)
Width: 4–6 in (10–15 cm)

Hind Print
Length: 9–12 in (23–30 cm)
Width: 5–7 in (13–18 cm)

Straddle
10–20 in (25–50 cm)

Stride
Walking: 24–40 in (60–100 cm)

Size (male>female)
Height: 3–3.5 ft (90–110 cm)
Length: 6–7 ft (1.8–2.1 m)

Weight
200–850 lb (90–390 kg)

*walking
(fast)*

GRIZZLY BEAR
(Brown Bear)
Ursus arctos

The magnificent and imposing Grizzly Bear symbolizes mountain wilderness for many of us, and Alaska has the highest and densest Grizzly population in North America. This bear prefers open country and valley bottoms. It is sensitive to human activity and in winter it enters a deep slumber, so few of its tracks are seen.

The Grizzly's huge prints each show four or five toes, with very long claws and a small heel pad on the fore print. A solid rear heel makes for a sturdy hind print. The toes are closely set in a line; the inner toe is the smallest. The usual walking track pattern shows the hind print registering in front of the fore print. A slower gait results in a trail like the Black Bear's (p. 36). The Grizzly occasionally gallops. Well-worn trails may lead to digs, trees with claw marks high up on the trunks or even a cached carcass. If you find a carcass, take care—the unpredictable Grizzly is probably nearby.

Similar Species: A Black Bear's prints are smaller, with shorter claw marks and toes arranged more in an arc, and its territorial tree-scratchings are lower on the trunk.

Polar Bear

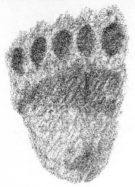

hind

walking (in snow)

Hind Print
(fore print is seldom seen)
Length: 12–13 in (30–33 cm)
Width: to 9 in (23 cm)

Straddle
to 20 in (50 cm)

Stride
Walking: 10–20 in (25–50 cm)

Size (male>female)
Height: 3–4 ft (90–120 cm)
Length: 6.5–7.5 ft (2–2.3 m)

Weight
600–1100 lb (270–500 kg)

POLAR BEAR
Ursus maritimus

The Polar Bear is truly the spirit of the far north. However, few of us will have the pleasure of finding this resilient bear's tracks, let alone the bear itself. Huge and massively furred, the Polar Bear roams the icy northern coast of Alaska and patrols the ice floes in search of seals or carrion.

The thick fur on a Polar Bear's foot keeps it warm, and obscures the finer details of the prints. Clear prints show five toes on each foot; the claws do not always register. When a Polar Bear walks, the hind print registers on or behind the smaller fore print. The Polar Bear's tracks are most likely to be found in snow, where its feet may drag; there is little room for confusion because there is little else in the far north that leaves such enormous tracks.

Similar Species: The smaller Black Bear (p. 36), with similar, smaller prints, does not visit the northern coast. The Grizzly Bear (p. 32), with longer claws, does.

Black Bear

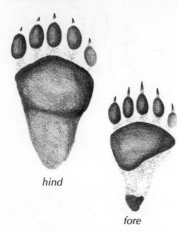

hind

fore

Fore Print
Length: 4–6.3 in (10–16 cm)
Width: 3.8–5.5 in (9.5–14 cm)

Hind Print
Length: 6–7 in (15–18 cm)
Width: 3.5–5.5 in (9–14 cm)

Straddle
9–15 in (23–38 cm)

Stride
Walking: 17–23 in (43–58 cm)

Size (male>female)
Height: 3–3.5 ft (90–110 cm)
Length: 5–6 ft (1.5–1.8 m)

Weight
200–600 lb (90–270 kg)

walking (slow)

36

BLACK BEAR
Ursus americanus

The Black Bear is widespread in forested areas throughout Alaska, but do not expect to see its tracks in winter, when it sleeps deeply. Finding fresh bear tracks can be a thrill, but take care; the bear may be just ahead. Never underestimate the potential power of a surprised bear!

Black Bear prints resemble smaller human prints, but wider and with claw marks. The small inner toe rarely registers. The forefoot's small heel pad often shows and the hind foot has a big heel. The bear's slow walk results in a slightly pigeon-toed double register of the hind print on the fore print. More frequently, at a faster pace, the hind foot oversteps the fore, as shown for the Grizzly Bear (p. 32). When a bear runs, the two hind feet register in front of the forefeet in an extended cluster. Along well-worn bear paths, look for 'digs'—patches of dug-up earth—and 'bear trees' whose scratched bark shows that these bears climb.

Similar Species: Grizzly Bear prints (often larger) show toes set more in a line and closer together, with longer claws.

Gray Wolf

fore

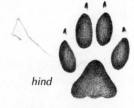

hind

Fore Print
(hind print is slightly smaller)
Length: 4–5.5 in (10–14 cm)
Width: 2.5–5 in (6.5–13 cm)

Straddle
3–7 in (7.5–18 cm)

Stride
Walking: 15–32 in (38–80 cm)
Galloping: 3 ft (90 cm)
    leaps to 9 ft (2.7 m)

Size (female is slightly smaller)
Height: 25–37 in (65–95 cm)
Length: 3.5–5.3 ft (1.1–1.6 m)

Weight
70–120 lb (32–55 kg)

walking   *trotting*

GRAY WOLF
(Timber Wolf)
Canis lupus

The soulful howl of the wolf epitomizes the outdoor experience, but few people ever hear it—your best bet is national parks or remote, undisturbed areas. The rarely seen Gray Wolf, the largest of the wild dogs, may travel in packs or alone.

A wolf leaves a straight alternating track pattern of large, oval prints that each show all four claws, with the smaller hind print registering directly on the fore print. The lobing on the fore and hind heel pads differs. In deep snow, wolves sensibly follow their leader, sometimes dragging their feet. When a wolf trots, notice how the hind print has a slight lead and falls to one side, giving an unbalanced appearance. Wolves and Coyotes (p. 40) gallop in the same way.

Similar Species: Domestic Dog (*Canis familiaris*) prints, rarely as large as a wolf's, fall in a haphazard pattern with a less direct register, and the inner toes tend to splay more. A Wolverine's (p. 56) print may show just four toes, but the pad shapes are very different.

Coyote

fore

hind

Fore Print
(hind print is slightly smaller)
Length: 2.4–3.2 in (6–8 cm)
Width: 1.6–2.4 in (4–6 cm)

Straddle
4–7 in (10–18 cm)

Stride
Walking: 8–16 in (20–40 cm)
Galloping: 2.5–10 ft (0.8 m–3 m)

Size (female is slightly smaller)
Height: 23–26 in (58–65 cm)
Length: 32–40 in (80–100 cm)

Weight
20–50 lb (9–23 kg)

walking   *gallop group*

COYOTE
(Brush Wolf, Prairie Wolf)
Canis latrans

This widespread and adaptable canine prefers open grasslands or woodlands. It hunts rodents and larger prey, either on its own, with a mate or in a family pack.

The oval fore prints are slightly larger than the hind prints. The fore heel pad differs from the hind heel pad, which rarely registers clearly. The two outer toes usually lack claw marks. A Coyote's tail hangs down, leaving a dragline in dust or sand. Coyotes typically walk or trot in an alternating pattern—the walk having a wider straddle. A Coyote's trotting trail is often very straight. When a Coyote gallops, the hind feet fall in front of its forefeet; the faster it goes, the straighter the gallop group.

Similar Species: A Domestic Dog's (*Canis familiaris*) less-oval prints splay more, and its trail is erratic. Foot hairs make Red Fox (p. 42) prints (usually smaller) less clear.

Red Fox

fore

hind

Fore Print
(hind print is slightly smaller)
Length: 2.1–3 in (5.3–7.5 cm)
Width: 1.6–2.3 in (4–5.8 cm)

Straddle
2–3.5 in (5–9 cm)

Stride
Walking: 12–18 in (30–45 cm)
Side-trotting:
    14–21 in (36–53 cm)

Size (vixen is slightly smaller)
Height: 14 in (35 cm)
Length: 22–25 in (55–65 cm)

Weight
7–15 lb (3.2–7 kg)

walking   *trotting*

RED FOX
Vulpes vulpes

   This beautiful and notoriously cunning fox, found throughout most of Alaska, prefers mountainous forests and open areas. It is a very adaptable and intelligent animal.

   The finer details of a Red Fox's tracks are obscured by the hair on its feet—only parts of the toes and heel pads show. The horizontal or slightly curved bar across the fore heel pad is diagnostic. A Red Fox leaves a distinctive, straight trail of alternating prints: the hind foot direct registers on the wider fore print. When a fox side-trots, the paired prints show the hind print falling to one side of the fore print in typical canid fashion. Foxes gallop like Coyotes (p. 40). The faster the gallop, the straighter the gallop group.

Similar Species: Domestic Dog (*Canis familiaris*) prints of similar size lack the bar on the heel pad. Arctic Fox (p. 44) prints, which are usually smaller, also lack the bar. Small Coyote prints are similar, but with a wider straddle.

Arctic Fox

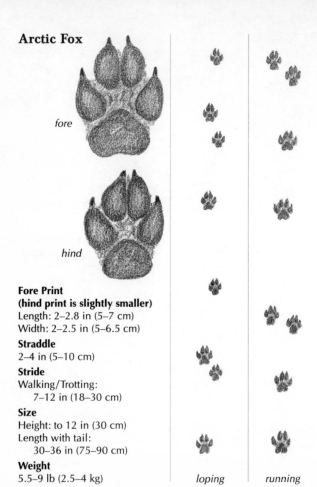

fore

hind

Fore Print
(hind print is slightly smaller)
Length: 2–2.8 in (5–7 cm)
Width: 2–2.5 in (5–6.5 cm)

Straddle
2–4 in (5–10 cm)

Stride
Walking/Trotting:
   7–12 in (18–30 cm)

Size
Height: to 12 in (30 cm)
Length with tail:
   30–36 in (75–90 cm)

Weight
5.5–9 lb (2.5–4 kg)

loping

running

ARCTIC FOX
Alopex lagopus

This delightful fox of the North changes from a thick and warm, white winter coat to a thinner, blue-brown one in summer. Although it is a solitary animal, it can be curious; in remote areas of northern Alaska it may come quite close to observe you.

Arctic Fox tracks are typical of dog-family tracks, with the hind prints slightly smaller than the fore prints. Both fore and hind prints show four toes, and the claws usually register. The close-set toes register quite clearly in summer. In winter, however, when thick hair covers the pads for warmth, print detail is obscured. When this fox walks or trots, it leaves a neat trail of alternating prints nearly in a line. If you come across an old bear kill, look around for fox tracks, because foxes frequently follow bears to feed on the leftovers.

Similar Species: The Red Fox (p. 42), with slightly smaller prints, has a distinctive heel-pad bar, but it may be obscured by hair.

45

Lynx

fore

hind

walking

Fore Print
(hind print is slightly smaller)
Length: 3.5–4.5 in (9–11 cm)
Width: 3.5–4.8 in (9–12 cm)

Straddle
6–9 in (15–23 cm)

Stride
Walking: 12–28 in (30–70 cm)

Size
Length: 2.5–3 ft (75–90 cm)

Weight
15–30 lb (7–14 kg)

LYNX
Lynx canadensis

This large cat would be a thrill to see, but, because it is sensitive to human interference, the elusive Lynx is abundant only in the dense forests of remote and undisturbed parts of Alaska. With its huge feet and relatively lightweight body, the Lynx stays on top of the snow as it pursues its main prey, the Snowshoe Hare (p. 66).

This cautious walker leaves an alternating track pattern with neat direct registers of the hind foot on top of the fore print. Thick fur on the feet often results in prints that are big, round depressions with no detail. In deeper snow, the print may be extended by 'handles' off to the rear. However deep the snow, this cat sinks no more than 8 inches (20 cm) and it rarely drags its feet. A Lynx is more likely to bound than to run. Its curious nature results in a meandering trail that may lead to a partially buried food cache.

Similar Species: No other large cat occurs in Alaska. Coyote (p. 40), Fox (pp. 42–45) and Domestic Dog (*Canis familiaris*) prints show claw marks, and their length exceeds their width.

47

Domestic Cat

fore

hind

Fore Print
(hind print is slightly smaller)
Length: 1–1.6 in (2.5–4 cm)
Width: 1–1.8 in (2.5–4.5 cm)

Straddle
2.4–4.5 in (6–11 cm)

Stride
Walking: 5–8 in (13–20 cm)
Loping/Galloping:
    14–32 in (35–80 cm)

Size (male>female)
Height: 20–22 in (50–55 cm)
Length with tail: 30 in (75 cm)

Weight
6.5–13 lb (3–6 kg)

walking

loping to
galloping

DOMESTIC CAT
(House Cat)
Felis catus

The tracks of the familiar and abundant Domestic Cat can show up almost any place where there are people. Abandoned cats may roam further afield; these 'feral cats' lead a pretty wild and independent existence. Domestic Cats can come in many shapes, sizes and colors.

As with all felines, a Domestic Cat's fore print and slightly smaller hind print both show four toe pads. Its retractable claws, kept clean and sharp for catching prey, do not register. Cat prints usually show a slight asymmetry, with one toe leading the others. A Domestic Cat makes a neat alternating walking track pattern, usually in direct register, as one would expect from this animal's fastidious nature. When a cat picks up speed, it leaves clusters of four prints, the hind feet registering in front of the forefeet.

Similar Species: The Lynx's (p. 46) much larger prints are obscured by fur on the sole of its feet. Fox (pp. 42–45) and Domestic Dog (*Canis familiaris*) prints show claw marks.

Harbor Seal

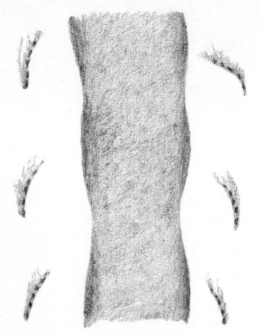

beach track

Size (male>female)
Length: 4–6 ft (1.2–1.8 m)
Weight
180–300 lb (80–140 kg)

HARBOR SEAL
Phoca vitulina

This common seal frequents isolated coastal beaches. One of the smaller seals, it is quite shy and will usually slide off its rocky sentry post into the sea to escape curious naturalists. Look in sandy and muddy areas between rocky platforms for its tracks, which are unmistakable in their location and large size. Sometimes a Harbor Seal works its way up a river, so you may find its tracks along riverbanks.

Seals, not the most graceful of animals on land, are unrivaled in the water. Their heavy, fat bodies and flipper-like feet can leave messy tracks: a wide trough (made by their cumbersome bellies) with dents alongside (made as the seals pushed themselves along with their forefeet). Look for the marks made by the nails.

Similar Species: The Northern Fur Seal (*Callorhinus ursinus*), seen on island shores and sometimes the mainland coast, can be twice as heavy. The Northern Sea Lion (*Eumetopias jubatus*) may weigh seven times as much as the little Harbor Seal. Other seals and sea lions can also be found on the remote offshore islands.

Sea Otter

fore

hind

Fore Print
Length: 2–3 in (5–7.5 cm)
Width: 2–3 in (5–7.5 cm)

Hind Print
Length: to 6 in (15 cm)
Width: to 6 in (15 cm)

Straddle
to 12 in (30 cm)

Stride
Walking: to 15 in (38 cm)

Size (male>female)
Length with tail: 2.5–6 ft (75–180 cm)

Weight
25–80 lb (11–36 kg)

walking

SEA OTTER
Enhydra lutris

As fun-loving as its freshwater cousin, the River Otter (p. 54), this large inhabitant of coastal waters is so suited to the sea that it seldom comes onto land. If you do catch sight of a Sea Otter on land, it might well be cavorting with sea lions, or waiting out a violent storm. With luck you may hear the tapping sound of an otter knocking on a shellfish with a stone to break it open. Sea Otter scat is recognizable by its crumbly texture and the fragments of shell in it.

Sea Otter tracks are most likely on beaches where kelp grows offshore. Like the seals, this otter is not as graceful on land as it is in the sea, because its webbed hind feet are almost flippers. The roundish fore print is much smaller than the hind print and the trail usually shows an alternating walking pattern.

Similar Species: The Harbor Seal (p. 50) also has flipper-like feet, but leaves an obvious trough with its belly and has much larger forefeet.

River Otter

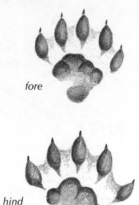

fore

hind

Fore Print
Length: 2.5–3.5 in (6.5–9 cm)
Width: 2–3 in (5–7.5 cm)

Hind Print
Length: 3–4 in (7.5–10 cm)
Width: 2.3–3.3 in (5.8–8.5 cm)

Straddle
4–9 in (10–23 cm)

Stride
Loping: 12–27 in (30–70 cm)

Size
(female is two-thirds the size of male)
Length with tail: 3–4.3 ft (90–130 cm)

Weight
10–25 lb (4.5–11 kg)

loping (fast)

RIVER OTTER
Lontra canadensis

No animal knows how to have more fun than a River Otter. If you are lucky enough to watch one at play, you will not soon forget the experience. Widespread and well-adapted for the aquatic environment, this otter lives along waterbodies. Expect to find a wealth of evidence along riverbanks in an otter's home territory. An otter in the forest is usually on its way to another waterbody.

In soft mud the River Otter's five-toed feet, especially the hind ones, register evidence of webbing. The inner toes are set slightly apart. If the forefoot's metacarpal pad registers, it lengthens the print. Very variable, otter trails usually show the typical mustelid 2x2 loping, but, with faster gaits, show groups of four and three prints. The thick, heavy tail often leaves a dragline. Otters love to slide, often down riverbanks, leaving troughs nearly 1 foot (30 cm) wide. In summer they roll and slide on grass and mud.

Similar Species: Mink (p. 60) prints are about half the size. Martens (p. 58), with similar-sized prints, are found in forested regions. Mink and Marten trails lack conspicuous tail drag.

55

Wolverine

fore

hind

Fore Print
Length with heel: 4–7.5 in (10–19 cm)
Width: 4–5 in (10–13 cm)

Hind Print
Length: 3.5–4 in (9–10 cm)
Width: 4–5 in (10–13 cm)

Straddle
7–9 in (18–23 cm)

Stride
Walking: 3–12 in (7.5–30 cm)
Running: 10–40 in (25–100 cm)

Size (female is slightly smaller)
Height: 16 in (40 cm)
Length: 32–46 in (80–120 cm)

Weight
18–45 lb (8–21 kg)

loping (slow)

WOLVERINE
Gulo gulo

The reputation of the robust and powerful Wolverine, largest of the mustelids, has earned it many nicknames, such as 'skunk bear' and 'Indian devil.' Wolverines live in coniferous mountain forests; their need for pristine wilderness has resulted in a scarce and scattered distribution throughout much of their range.

As with other mustelids, Wolverines have five toes, but the inner toe rarely registers in the print. Though the forefoot registers a small heel pad, the hind foot rarely does. Because of its low, squat shape, the Wolverine leaves a host of erratic typical mustelid trails: an alternating walking pattern, the typical 2x2 loping pattern with its print pairs and the common loping pattern of three- and four-print groups.

Similar Species: Most other mustelids (pp. 58–65) make much smaller prints and their track patterns are less erratic. When only four toes register, Wolverine tracks may be mistaken for Gray Wolf (p. 38) or Domestic Dog (*Canis familiaris*) tracks, but the pad shapes are quite different.

Marten

Fore and Hind Prints
Length: 1.8–2.5 in (4.5–6.5 cm)
Width: 1.5–2.8 in (3.8–7 cm)

Straddle
2.5–4 in (6.5–10 cm)

Stride
Walking: 4–9 in (10–23 cm)
2x2 loping:
9–46 in (23–120 cm)

Size (male>female)
Length with tail:
21–28 in (53–70 cm)

Weight
1.5–2.8 lb (0.7–1.3 kg)

walking   *2x2 loping*

MARTEN (American Sable)
Martes americana

This aggressive predator is found in coniferous and mixed-wood forests throughout Alaska.

The Marten seldom leaves a clear print: often just four toes register, the heel pad is undeveloped and, in winter, the hairiness of the feet often obscures all pad detail, especially the poorly developed palm pads. In the Marten's alternating walking pattern, the hind foot registers on the fore print. In 2x2 loping, the hind prints fall on the fore prints to form slightly angled print pairs, in a typical mustelid pattern. Loping patterns may appear as three- or four-print clusters (see the River Otter, p. 54). Follow the criss-crossing trails—if the Marten has scrambled up a tree, look for the sitzmark where it has jumped down again.

Similar Species: Male Mink (p. 60) prints overlap in size with small female Marten prints but Mink, usually near water, do not climb trees.

Mink

fore

hind

Fore and Hind Prints
Length: 1.3–2 in (3.3–5 cm)
Width: 1.3–1.8 in (3.3–4.5 cm)
Straddle
2.1–3.5 in (5.3–9 cm)
Stride
Walking/2x2 loping: 8–36 in (20–90 cm)
Size (male>female)
Length with tail: 19–28 in (48–70 cm)
Weight
1.5–3.5 lb (0.7–1.6 kg)

2x2 loping

MINK
Mustela vison

The lustrous Mink, widespread throughout the state, prefers watery habitats surrounded by brush or forest. At home as much on land as in water, this nocturnal hunter can be exciting to track. Like the River Otter (p. 54), the Mink slides in snow, carving out a trough up to 6 inches (15 cm) wide for an observant tracker to spot.

The Mink's fore print shows five (perhaps four) toes, with five loosely connected palm pads in an arc, but the hind print shows only four palm pads. The metacarpal pad of the forefoot rarely registers, but the furred heel of the hind foot may register, lengthening the hind print. The Mink prefers the typical mustelid 2x2 loping, making consistently spaced, slightly angled double prints. Its diverse track patterns also include alternating walking; loping with three- and four-print groups (like the River Otter); and bounding, like hares (p. 66).

Similar Species: Small Martens (p. 58), with similar prints, do not have a consistent 2x2 loping gait or live near water. Weasels (pp. 62–65) make similar, smaller tracks.

Short-tailed Weasel

Fore and Hind Prints
Length: 0.8–1.3 in (2–3.3 cm)
Width: 0.5–0.6 in (1.3–1.5 cm)

Straddle
1–2.1 in (2.5–5.3 cm)

Stride
2x2 loping: 9–36 in (23–90 cm)

Size (male>female)
Length with tail: 8–14 in (20–35 cm)

Weight
1–6 oz (30–170 g)

2x2 loping

SHORT-TAILED WEASEL
(Ermine, Stoat)
Mustela erminea

Weasels are active year-round hunters, with an avid appetite for rodents. Following their tracks can reveal much about these nimble creatures' activities. Tracks are most evident in winter, when they frequently burrow into the snow or pursue rodents into their holes. Some weasel trails may lead you up a tree. Weasels sometimes take to water.

The Short-tailed Weasel is widely distributed throughout Alaska. It prefers woodlands and meadows up to higher elevations but does not favor wetlands or dense coniferous forests. Its 2x2 loping tracks may fall in clusters, with alternating short and long strides.

Similar Species: Large male Least Weasel (p. 64) tracks may be the same size as a small female Short-tailed's.

Least Weasel

Fore and Hind Prints
Length: 0.5–0.8 in (1.3–2 cm)
Width: 0.4–0.5 in (1–1.3 cm)

Straddle
0.8–1.5 in (2–3.8 cm)

Stride
2x2 loping: 5–20 in (13–50 cm)

Size (male>female)
Length with tail:
    6.5–9 in (17–23 cm)

Weight
1.3–2.3 oz (35–65 g)

2x2 loping

LEAST WEASEL
Mustela nivalis

The typical weasel gait is a 2x2 lope, leaving a trail of paired prints. Because of a weasel's light weight and small, hairy feet, the pad detail is often unclear, especially in snow. Even with clear tracks, the inner (fifth) toe rarely registers. To identify the weasel species, pay close attention to straddle and stride, but note that small females of a larger species and large males of a smaller species may have similar tracks. Also check the habit displayed in loping patterns and note the distribution and habitat.

The Least Weasel, found throughout Alaska, is the smallest weasel, with the least-clear tracks. Tracks may be found around wetlands and in open woodlands and fields.

Similar Species: A small female Short-tailed Weasel's (p. 62) tracks may resemble a large male Least Weasel's, but Short-tailed Weasels do not frequent wet areas, preferring upland areas and woodlands.

Snowshoe Hare

hind

fore

Fore Print
Length: 2–3 in (5–7.5 cm)
Width: 1.5–2 in (3.8–5 cm)
Hind Print
Length: 4–6 in (10–15 cm)
Width: 2–3.5 in (5–9 cm)
Straddle
6–8 in (15–20 cm)
Stride
Hopping: 0.8–4.3 ft (25–130 cm)
Size
Length: 12–21 in (30–53 cm)
Weight
2–4 lb (0.9–1.8 kg)

hopping

SNOWSHOE HARE
(Varying Hare)
Lepus americanus

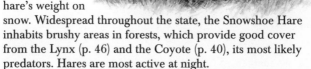

This hare is well known for its color change from summer brown to winter white, and for the huge hind feet (much larger than the forefeet) that enable it to 'float' on top of snow. In winter, heavy fur on the hind feet thickens the toes, which can splay out to further distribute the hare's weight on snow. Widespread throughout the state, the Snowshoe Hare inhabits brushy areas in forests, which provide good cover from the Lynx (p. 46) and the Coyote (p. 40), its most likely predators. Hares are most active at night.

The Snowshoe Hare's most common track pattern is a hopping one, with triangular four-print groups; they can be quite long if the hare moves quickly. Hares make well-worn runways that are often used as escape runs. You may encounter a resting hare, since hares do not live in burrows. Twigs and stems neatly cut at a 45° angle also indicate this hare's presence.

Similar Species: The Alaska Hare (*Lepus othus*) leaves larger tracks in northern and northwestern coastal areas.

Collared Pika

fore

hind

Fore Print
Length: 0.8 in (2 cm)
Width: 0.6 in (1.5 cm)

Hind Print
Length: 1–1.2 in (2.5–3 cm)
Width: 0.6–0.8 in (1.5–2 cm)

Straddle
2.5–3.5 in (6.5–9 cm)

Stride
Walking/Bounding: 4–10 in (10–25 cm)

Size
Length: 6.5–8.5 in (17–22 cm)

Weight
4–6 oz (110–170 g)

bounding

COLLARED PIKA
Ochotona collaris

When hiking high in the mountains, you are more likely to hear the squeak of this cousin of the rabbit than to see it, because it is quick to disappear under the rocks for protection. Confined to areas of high elevation, the Collared Pika rarely leaves good tracks: in summer it usually travels across exposed, rocky areas and in winter it stays under the snow and feeds on stored food. Pika tracks are best sought in spring, on patches of snow or mud.

The fore print usually shows five toes (sometimes four), but the hind print always shows four. The prints may appear in an erratic alternating pattern or in three- and four-print bounding groups. Little hay piles, left to dry in the sun in preparation for the long winter ahead, are a conspicuous sign of the Collared Pika's presence.

Similar Species: Because of this pika's high-mountain habitat, its prints are rarely confused with a hare's (p. 66).

Porcupine

fore

hind

Fore Print
Length: 2.3–3.3 in (5.8–8.5 cm)
Width: 1.3–1.9 in (3.3–4.8 cm)

Hind Print
Length: 2.8–4 in (7–10 cm)
Width: 1.5–2 in (3.8–5 cm)

Straddle
5.5–9 in (14–23 cm)

Stride
Walking: 5–10 in (13–25 cm)

Size
Length with tail: 25–40 in (65–100 cm)

Weight
10–28 lb (4.5–13 kg)

walking

PORCUPINE
Erethizon dorsatum

This notorious rodent rarely runs— its many long quills are a formidable defense. Widespread in Alaska, the Porcupine shows a preference for forests but can also be seen in more-open areas.

The Porcupine's preferred pigeon-toed waddling gait leaves an alternating walking pattern, with the hind print registering on or slightly in front of the shorter fore print. Look for long claw marks on all prints. The fore print shows four toes and the hind print five. On clear prints, the unusual pebbly surface of the solid heel pads may show, but a Porcupine's tracks are often obscured by scratches from its heavy, spiny tail. In deeper snow, this squat animal drags its feet, and it may leave a trough with its body. A Porcupine's trail might lead you to a tree, where these animals spend much of their time feeding; look for chewed bark or nipped twigs on the ground.

Similar Species: No animal in Alaska makes comparable tracks.

Beaver

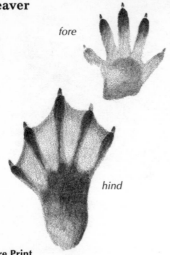

fore

hind

Fore Print
Length: 2.5–4 in (6.5–10 cm)
Width: 2–3.5 in (5–9 cm)

Hind Print
Length: 5–7 in (13–18 cm)
Width: 3.3–5.3 in (8.5–13 cm)

Straddle
6–11 in (15–28 cm)

Stride
Walking: 3–6.5 in (7.5–17 cm)

Size
Length with tail:
    3–4 ft (90–120 cm)

Weight
28–75 lb (13–34 kg)

walking

BEAVER
Castor canadensis

Few animals leave as many signs of their presence as the Beaver, the largest North American rodent and a common sight around waterbodies throughout Alaska. Look for the conspicuous dams and lodges—capable of changing the local landscape—and the stumps of felled trees. Check trunks gnawed clean of bark for marks of the Beaver's huge incisors. A scent mound marked with castoreum, a strong-smelling yellowish fluid that Beavers produce, also indicates recent activity.

The thick, scaly tail of the Beaver may mar its tracks, as can the branches that it drags about for construction and food. Check the large hind prints for signs of webbing and broad toenails—the nail of the second inner toe usually does not show. Rarely do all five toes on each foot register. Irregular foot placement in the alternating walking gait may produce a direct or a double register. Repeated path use results in well-worn trails.

Similar Species: The Beaver's many signs, including large hind prints, minimize confusion. Muskrat (p. 74) prints are smaller.

Muskrat

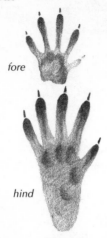

fore

hind

Fore Print
Length: 1.1–1.5 in (2.8–3.8 cm)
Width: 1.1–1.5 in (2.8–3.8 cm)

Hind Print
Length: 1.6–3.2 in (4–8 cm)
Width: 1.5–2.1 in (3.8–5.3 cm)

Straddle
3–5 in (7.5–13 cm)

Stride
Walking: 3–5 in (7.5–13 cm)
Running: to 1 ft (30 cm)

Size
Length with tail: 16–25 in (40–65 cm)

Weight
2–4 lb (0.9–1.8 kg)

walking

MUSKRAT
Ondatra zibethicus

Like the Beaver, this rodent is found throughout Alaska, wherever there is water. Beavers (p. 72) are very tolerant of Muskrats and even allow them to live in parts of their lodges. Active all year, Muskrats leave plenty of signs. They dig extensive networks of burrows, often undermining the riverbank, so do not be surprised if you suddenly fall into a hidden hole! Also look for small lodges in the water and beds of vegetation on which Muskrats rest, sun and feed in summer.

The small inner toe of the five on each forefoot rarely registers. The hind print shows five well-formed toes that may have a 'shelf' around them, created by stiff hairs that aid in swimming. The common alternating walking trail shows track pairs that alternate from side to side, with the hind print just behind or slightly overlapping the fore print. In snow, a Muskrat's feet drag and its tail leaves a sweeping dragline.

Similar Species: Few animals share this water-loving rodent's habits. Beavers make larger prints and leave many other signs.

Hoary Marmot

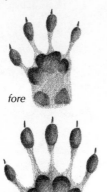

fore

hind

Fore and Hind Prints
Length: 1.8–2.8 in (4.5–7 cm)
Width: 1–2 in (2.5–5 cm)

Straddle
3.3–6 in (8.5–15 cm)

Stride
Walking: 2–6 in (5–15 cm)
Running: 6–14 in (15–35 cm)

Size (male>female)
Length with tail:
    17–32 in (43–80 cm)

Weight
5–15 lb (2.3–7 kg)

walking   *running*

HOARY MARMOT
Marmota caligata

These endearing squirrels seem to lead the good life, sleeping all winter and sunbathing on rocks in summer. Hoary Marmots live in small colonies in the high rocky areas of Alaska and dig extensive networks of burrows. They are a joy to watch as they play-fight.

The fore print shows four toes and three palm pads; two heel pads may also be evident. The hind print shows five toes, four palm pads and two poorly registering heel pads. In a marmot's usual alternating walking pattern, the hind foot registers over the fore print but, when a marmot runs, it leaves groups of four prints, with the two hind prints in front of the fore prints. This marmot's preference for rocky habitats makes its tracks hard to find, so look after spring or fall snowfalls.

Similar Species: The Alaska Marmot (*Marmota broweri*) of the Brooks Range has similar tracks. The Woodchuck (p. 78) makes smaller tracks in or near open woodlands.

Woodchuck

fore

hind

Fore and Hind Prints
Length: 1.8–2.8 in (4.5–7 cm)
Width: 1–2 in (2.5–5 cm)

Straddle
3.3–6 in (8.5–15 cm)

Stride
Walking: 2–6 in (5–15 cm)
Running: 6–14 in (15–35 cm)

Size (male>female)
Length with tail:
    20–25 in (50–65 cm)

Weight
5.5–12 lb (2.5–5.4 kg)

walking   *running*

78

WOODCHUCK
(Whistle Pig, Groundhog, Marmot)

Marmota monax

This robust member of the squirrel family is a common sight in open woodlands and adjacent open areas throughout Alaska. Always on the watch for predators, but not too troubled by humans, the Woodchuck never wanders far from its burrow. This marmot hibernates during winter but emerges in early spring; look for tracks in late spring snowfalls and in the mud around the burrow entrances.

A Woodchuck's fore print shows four toes, three palm pads and two heel pads, though the heel pads are not always evident. The hind print shows five toes, four palm pads and two poorly registering heel pads. Woodchucks usually leave an alternating walking pattern, with the hind print registering over the fore print. When a Woodchuck runs from danger, it makes groups of four prints, hind ahead of fore.

Similar Species: The Hoary Marmot (p. 76) makes similar, but larger, tracks in rocky, mountainous regions.

Arctic Ground Squirrel

fore

hind

Fore Print
Length: 1–1.3 in (2.5–3.3 cm)
Width: 0.5–1 in (1.3–2.5 cm)

Hind Print
Length: 1.1–1.5 in (2.8–3.8 cm)
Width: 0.8–1.3 in (2–3.3 cm)

Straddle
2.3–4 in (5.8–10 cm)

Stride
Running: 7–20 in (18–50 cm)

Size
Length with tail: 8–13 in (20–33 cm)

Weight
6–10 oz (170–280 g)

running

ARCTIC GROUND SQUIRREL

Spermophilus parryii

Widespread throughout Alaska, the Arctic Ground Squirrel enjoys a variety of different habitats, from grassy meadows to subalpine areas. These bold squirrels can be very tame when they beg food, but, no matter how sweet they are, don't give in—our unhealthy diet certainly doesn't suit them!

This animal's tracks may be evident near its many burrow entrances, in mud or in late or early snowfalls (it hibernates in winter). The hind print shows five toes but the smaller fore print usually shows four, and the two heel pads sometimes show. Both forefeet and hind feet have long claws that frequently register. Usually seen scurrying around, ground squirrels leave a typical squirrel track, with the hind print registering ahead of the fore print, which are generally placed diagonally.

Similar Species: No other ground squirrel lives in Alaska. Red Squirrels (p. 82) leave similar tracks in a more square-shaped running group, but only in forested areas.

Red Squirrel

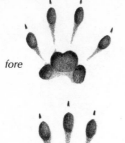

fore

hind

Fore Print
Length: 0.8–1.5 in (2–3.8 cm)
Width: 0.5–1 in (1.3–2.5 cm)

Hind Print
Length: 1.5–2.3 in (3.8–5.8 cm)
Width: 0.8–1.3 in (2–3.3 cm)

Straddle
3–4.5 in (7.5–11 cm)

Stride
Running: 8–30 in (20–75 cm)

Size
Length with tail:
    9–15 in (23–38 cm)

Weight
2–9 oz (55–260 g)

bounding

bounding
(in deep snow)

RED SQUIRREL (Pine Squirrel, Chickaree)
Tamiasciurus hudsonicus

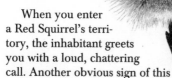

When you enter a Red Squirrel's territory, the inhabitant greets you with a loud, chattering call. Another obvious sign of this forest dweller, which is found throughout most of Alaska, is its large middens—piles of cone scales and cores left beneath trees—that indicate favorite squirrel-feeding sites.

Active year-round in their small territories, Red Squirrels leave an abundance of trails that lead from tree to tree or down a burrow. These energetic animals mostly bound. This gait leaves groups of four prints, with the hind prints falling in front of the fore prints, which tend to be side by side (but not always). Four toes show on each fore print, and five on each hind print. The heels often do not register when squirrels move quickly. In deeper snow, the prints merge to form pairs of diamond-shaped tracks.

Similar Species: Northern Flying Squirrel (p. 84) tracks show a similar pattern, but have a smaller straddle and smaller prints.

Northern Flying Squirrel

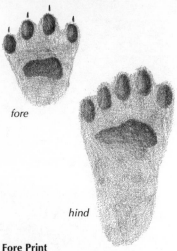

fore

hind

Fore Print
Length: 0.5–0.8 in (1.3–2 cm)
Width: 0.5 in (1.3 cm)
Hind Print
Length: 1.3–1.8 in (3.3–4.5 cm)
Width: 0.8 in (2 cm)
Straddle
3–3.8 in (7.5–9.5 cm)
Stride
Running: 11–30 in (28–75 cm)
Size
Length with tail: 9–12 in (23–30 cm)
Weight
4–6.5 oz (110–180 g)

*sitzmark
into bounding*

NORTHERN FLYING SQUIRREL
Glaucomys sabrinus

This acrobat can be found in the coniferous forests of southern Alaska. It prefers widely spaced forests, where it can glide from tree to tree by night, using the membranous flap between its legs. Northern Flying Squirrels will den up together in a tree cavity for warmth.

Because of its gliding, this squirrel does not leave as many tracks as other squirrels. Evidence is scarce in summer but in winter you may come across a sitzmark—the distinctive pattern that an animal leaves when it lands in the snow—and a short bounding trail, made as it rushes off to the nearest tree or to do some quick foraging. The bounding track pattern is typical of squirrels and other rodents, but with the hind feet registering only slightly in front of the fore prints–though often all four feet register in a row.

Similar Species: Red Squirrels (p. 82) usually make larger prints and rarely leave a sitzmark, but unclear tracks in deep snow can be impossible to distinguish from a flying squirrel's.

Norway Rat

fore

hind

Fore Print
Length: 0.7–0.8 in (1.8–2 cm)
Width: 0.5–0.7 in (1.3–1.8 cm)

Hind Print
Length: 1–1.3 in (2.5–3.3 cm)
Width: 0.8–1 in (2–2.5 cm)

Straddle
2–3 in (5–7.5 cm)

Stride
Walking: 1.5–3.5 in (3.8–9 cm)
Bounding: 9–20 in (23–50 cm)

Size
Length with tail:
    13–19 in (33–48 cm)

Weight
7–18 oz (200–510 g)

walking

NORWAY RAT
Rattus norvegicus

Active both day and night, this despised rat is wide-spread almost anywhere that humans have decided to build homes. Not entirely dependent on people, it may live in the wild as well.

The fore print shows four toes and the hind print five. When it bounds, this colonial rat leaves four-print groups, with the hind prints in front of the diagonally placed fore prints. Sometimes one of the hind feet direct registers on a fore print, creating a three-print group. More commonly, rats leave an alternating walking pattern, with the larger hind prints close to or overlapping the fore prints; the hind heel does not show. In snow, the tail often leaves a dragline. Rats live in groups, so you may find many trails together, often leading to their 5-cm (2-inch) wide burrows.

Similar Species: Mice (pp. 90–93) prints are much smaller. Red Squirrel (p. 82) tracks show distinctive squirrel characteristics.

Meadow Vole

fore

hind

Fore Print
Length: 0.3–0.5 in (0.8–1.3 cm)
Width: 0.3–0.5 in (0.8–1.3 cm)

Hind Print
Length: 0.3–0.6 in (0.8–1.5 cm)
Width: 0.3–0.6 in (0.8–1.5 cm)

Straddle
1.3–2 in (3.3–5 cm)

Stride
Walking/Trotting:
    1.3–3 in (3.3–7.5 cm)
Bounding: 4–8 in (10–20 cm)

Size
Length with tail:
    5.5–8 in (14–20 cm)

Weight
0.5–2.5 oz (14–70 g)

walking

*bounding
(in snow)*

MEADOW VOLE
(Field Mouse)
Microtus pennsylvanicus

With so many vole species in Alaska, positive trail identification is next to impossible. Note, though, that the Meadow Vole commonly inhabits damp or wet habitats across most of the state.

When clear, which is seldom, vole fore prints show four toes while hind prints show five. A vole's walk and trot both leave a paired alternating track pattern; the hind foot occasionally direct registers on the fore print. Voles usually opt for a faster bounding; in the resulting print pairs, the hind print registers on the fore print. These voles lope quickly across open areas, creating a three-print track pattern. In winter, voles stay under the snow; when it melts, look for distinctive piles of cut grass from their ground nests. The bark at the bases of shrubs may show tiny teeth marks left by gnawing. In summer, well-used vole paths appear as little runways in the grass.

Similar Species: Other common voles species, with similar prints, are the Northern Red-backed Vole (*Cletherionomys rutilus*), the Singing Vole (*M. miurus*) and the Tundra Vole (*M. oeconomus*).

Keen's Mouse

bounding group

Fore Print
Length: 0.3–0.4 in (0.8 cm)
Width: 0.3 in (0.8 cm)
Hind Print
Length: 0.6 in (1.5 cm)
Width: 0.4 in (1 cm)
Straddle
1.4–1.8 in (3.6–4.5 cm)
Stride
Running: 2–5 in (5–13 cm)
Size
Length with tail:
    8–9 in (20–23 cm)
Weight
0.9–1.4 oz (25–40 g)

bounding

*bounding
(in snow)*

KEEN'S MOUSE
(Forest Deer Mouse, Sitka Mouse)

Peromyscus keeni
(formerly *P. sitkensis*)

The highly adaptable Keen's Mouse lives anywhere from arid valleys to alpine meadows, but is abundant only on Alaska's southern islands. It may enter buildings and stay active in winter.

The fore prints each show four toes, three palm pads and two heel pads. The hind prints each show five toes and three palm pads; the heel pads rarely register. It takes perfect, soft mud to get clear prints from such a tiny mammal. Bounding tracks, most noticeable in snow, show the hind prints falling in front of the fore prints. In soft snow the prints may merge to look like larger pairs of prints, with tail drag evident. A trail may lead up a tree or down into a burrow.

Similar Species: Many less common species of mice make identical tracks. The closely related Deer Mouse (*P. maniculatus*) may occur in parts of southern Alaska. The House Mouse (*Mus musculus*), associated more with humans, has very similar tracks. Voles (p. 88) have a longer stride. Jumping mouse (p. 92) prints may be similar in size. Shrews (p. 94) have a narrower straddle.

Meadow Jumping Mouse

*bounding
group*

Fore Print
Length: 0.3–0.5 in (0.8–1.3 cm)
Width: 0.3–0.5 in (0.8–1.3 cm)

Hind Print
Length: 0.5–1.3 in (1.3–3.3 cm)
Width: 0.5–0.7 in (1.3–1.8 cm)

Straddle
1.8–1.9 in (4.5–4.8 cm)

Stride
Bounding: 7–18 in (18–45 cm)
In alarm: 3–6 ft (90–180 cm)

Size
Length with tail: 7–9 in (18–23 cm)

Weight
0.6–1.3 oz (17–35 g)

bounding

MEADOW JUMPING MOUSE
Zapus hudsonius

Congratulations if you find and successfully identify the tracks of a Meadow Jumping Mouse! Though it lives throughout southern Alaska, its preference for grassy meadows—and its long, deep winter hibernation (about six months!)—makes locating tracks very difficult.

Jumping mouse tracks are distinctive if you do find them. The two smaller forefeet register between the long hind prints; the long heels do not always register and some prints show just the three long middle toes. The toes on the forefeet may splay so much that the side toes point backward. When bounding, these mice make short leaps. The tail may leave a dragline in soft mud or unseasonable snow. Clusters of cut grass stems, about 5 inches (13 cm) long, lying in meadows, are a more abundant sign of this rodent.

Similar Species: If the long hind heels do not register well, jumping mice prints may be mistaken for vole (p. 88) prints. They may look so abstract that they are mistaken for a small bird's (p. 116) or an amphibian's (pp. 118–123).

Dusky Shrew

bounding group

Fore Print
Length: 0.2 in (0.5 cm)
Width: 0.2 in (0.5 cm)
Hind Print
Length: 0.6 in (1.5 cm)
Width: 0.3 in (0.8 cm)
Straddle
0.8–1.3 in (2–3.3 cm)
Stride
Running: 1.2–2 in (3–5 cm)
Size
Length with tail: 3–5 in (7.5–13 cm)
Weight
0.1–1 oz (3–30 g)

bounding

DUSKY SHREW
Sorex monticolus

Though many species of tiny, frenetic shrews are found in Alaska, the most likely candidate is the widespread and adaptable Dusky Shrew. This shrew is just as happy in the high heathlands as in the lowland swamps. Its rapid activity makes it difficult to observe closely.

In its energetic and unending quest for food, a shrew usually leaves a four-print bounding pattern, but it may slow to an alternating walking pattern. The individual prints in a group are often indistinct but, in mud or shallow, wet snow, you can even count the five toes on each print. In deeper snow, a shrew's tail often leaves a dragline. If a shrew tunnels under the snow, it may leave a ridge of snow on the surface. A shrew's trail may go down a burrow.

Similar Species: The Pygmy Shrew (*S. hoyi*) is smaller. The larger Water Shrew (*S. palustris*) is often found by cold mountain streams. The Masked (Common) Shrew (*S. cinereus*) is common but more secretive.

Sasquatch

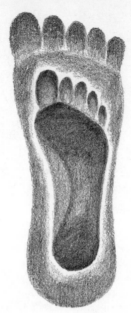

Foot Print
Length: average 14–17 in (35–43 cm)
Width: to 7 in (18 cm)

Hand Print
Length: to 1 ft (30 cm)
Width: to 7 in (18 cm)

Stride
Walking: 2–3 ft (60–90 cm)

Size (male>female)
Height: 6–8 ft (1.8–2.4 m)

Weight
400–1000 lb (180–450 kg)

SASQUATCH
(Bigfoot)
Homo cascadensis

Imagine the thrill of being one of the few people to find the tracks of the reclusive Sasquatch!

The enormous foot of this elusive, human-like inhabitant of the remote wilderness is much broader and flatter than a human's (*Homo sapiens*)—notice the raised arch on a human print. Its track patterns are like a human's. Most Sasquatch tracks have been found along rivers, but a clear print might be seen on firmer surfaces. If you find a naked print of huge dimensions, let somebody know!

Similar Species: Huge bear (pp. 32–37) prints have prominent claws and many other distinctions. Large human footprints are similar to a juvenile Sasquatch's, but naked human prints are mostly seen on sandy beaches near water, where (in summer) humans can be spotted lying around in large numbers, like lazy sea lions. Elsewhere the human foot is usually shod to protect it from sticks and stones. Also, irresponsible humans leave plenty of evidence of their presence on trails. If a raven has not beaten you to it, do the beautiful mountain wilderness a big favor and pick up the garbage.

BIRDS AND AMPHIBIANS

A guide to the animal tracks of Alaska is not complete without some consideration of the birds and amphibians found in the state.

Birds come in many shapes and sizes. Several species have been chosen to represent the main types common to Alaska, but the differences between bird species are not always reflected in their tracks.

Bird tracks can often be found in abundance in snow and are clearest in shallow, wet snow. The shores of streams and lakes are very reliable locations in which to find bird tracks—the mud there can hold a clear print for a long time. The sheer number of tracks made by shorebirds and waterfowl can be astonishing. Though some bird species prefer to perch in trees or soar across the sky, it can be entertaining to follow the tracks of those that do spend a lot of time on the ground. They can spin around in circles and lead you in all directions. The trail may suddenly end as the bird takes flight, or it might terminate in a pile of feathers, the bird having fallen victim to a hungry predator.

Many amphibians depend on moist environments, so look in the soft mud along the shores of lakes and ponds for their distinctive tracks. Though you may be able to distinguish frog tracks from toad tracks, since they generally move differently, it can be very difficult to identify the species.

Canada Goose

Print
Length: 4–5 in (10–13 cm)
Straddle
5–7 in (13–18 cm)
Stride
Walking: 5–7 in (13–18 cm)
Size
2.7–4 ft (80–120 cm)

CANADA GOOSE
Branta canadensis

This common goose is a familiar sight in open areas by lakes and ponds. Its huge, webbed feet leave prints that can often be seen in abundance along the muddy shores of just about any waterbody, including those in urban parks, where the Canada Goose's green-and-white droppings can accumulate in prolific amounts.

The webbed feet each have three long toes, all facing forward. These toes register well, but the webbing between them does not always show on the print. The prints point inward, which gives the bird a pigeon-toed appearance and may account for its waddling gait.

Similar Species: Many waterfowl, such as ducks and gulls (p. 102), leave similar, but usually smaller, prints. Exceptionally large tracks are likely a swan's (*Cygnus* spp.).

Herring Gull

Print
Length: 3.5 in (9 cm)
Straddle
4–6 in (10–15 cm)
Stride
4.5 in (11 cm)
Size
Length: 23–25 in (58–65 cm)

HERRING GULL
Larus argentatus

Herring gulls, with their long wings and webbed toes, are strong long-distance fliers as well as excellent swimmers. They are found in eastern parts of Alaska, and are concentrated in great numbers on the coast and near bodies of fresh water. Their slightly asymmetrical tracks show three toes. They have claws that register outside the webbing and the clawmarks are usually attached to the footprint. Most gulls have quite a swagger to their gait, and they leave a trail with the tracks turned strongly inward.

Similar Species: Gull species cannot be reliably identified by track alone, but smaller species have conspicuously smaller tracks. Duck tracks are difficult to distinguish from gull tracks. Canada Goose (p. 100) and swan (*Cygnus* spp.) tracks are usually larger.

Great Blue Heron

Print
Length: to 6.5 in (17 cm)
Straddle
8 in (20 cm)
Stride
9 in (23 cm)
Size
4.2–4.5 ft (1.3–1.4 m)

GREAT BLUE HERON
Ardea herodias

The refined and graceful image of this large heron symbolizes the precious wetlands in which it patiently hunts for food. Usually still and statuesque as it waits for a meal to swim by, this heron will have cause to walk from time to time, perhaps to find a better hunting location. Look for its large, slender tracks along the banks or mudflats of waterbodies in southern Alaska.

Not surprisingly, a bird that lives and hunts with such precision walks in a similar fashion, leaving straight tracks that fall in a nearly straight line. Look for the slender rear toe in the print.

Similar Species: Cranes (*Grus* spp.), widespread in similar habitats in Alaska, have similar, although possibly larger, tracks, but their smaller hind toes do not register.

Common Snipe

Print
Length: 1.5 in (3.8 cm)

Straddle
to 1.8 in (4.5 cm)

Stride
to 1.3 in (3.3 cm)

Size
11–12 in (28–30 cm)

COMMON SNIPE
Gallinago gallinago

This short-legged character is a resident of marshes and bogs, where its neat prints can often be seen in mud. Snipes are quite secretive when on the ground, and so you may be surprised if one suddenly flushes out from beneath your feet. Watch out for the occasional snipe perched on a snag or fencepost, and listen for the eerie whistle as a snipe dives from the sky.

The Common Snipe's neat prints show four toes, including a small rear toe that points inward. The bird's short legs and stocky body give it a very short stride.

Similar Species: Many shorebirds, including the Spotted Sandpiper (p. 108), leave similar tracks.

Spotted Sandpiper

Print
Length: 0.8–1.3 in (2–3.3 cm)

Straddle
to 1.5 in (3.8 cm)

Stride
Erratic

Size
7–8 in (18–20 cm)

SPOTTED SANDPIPER
Actitis macularia

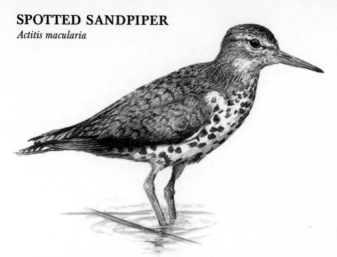

The bobbing tail of the Spotted Sandpiper is a common sight on the shores of lakes, rivers and streams, but you will usually find just one of these territorial birds in any given location. Because of its excellent camouflage, likely the first you will see of this bird will be when it flies away, its fluttering wings close to the surface of the water.

As it teeters up and down on the shore, it leaves trails of three-toed prints. Its fourth toe is very small and faces off to one side at an angle. Sandpiper tracks can have an erratic stride.

Similar Species: All sandpipers and plovers, including the Killdeer (*Charadrius vociferus*) leave similar tracks, although there is much diversity in size. The Common Snipe (p. 106) makes larger tracks.

Ruffed Grouse

Print
Length: 2–3 in (5–7.5 cm)
Straddle
2–3 in (5–7.5 cm)
Stride
Walking: 3–6 in (7.5–15 cm)
Size
15–19 in (34–48 cm)

RUFFED GROUSE
Bonasa umbellus

This ground-dweller prefers the quiet seclusion of coniferous forests in winter, so that will be the best place to find its tracks. If you follow them quietly, you may be startled when the grouse bursts from cover underneath your feet. Its excellent camouflage usually affords it good protection.

The three thick front toes leave very clear impressions, but the short rear toe, which is angled off to one side, does not always show up so well. This bird's neat, straight trail appears to reflect its cautious approach to life on the forest floor.

Similar Species: Other grouse and the ptarmigans (*Lagopus* spp.) leave similar tracks, but their prints may be enlarged and obscured by the winter feathers that they grow on their feet.

Great Horned Owl

Strike
Width: to 3 ft (90 cm)
Size
22 in (55 cm)

GREAT HORNED OWL
Bubo virginianus

Often seen resting quietly in trees by day, this wide-ranging owl prefers to hunt at night. An accomplished hunter in snow, the owl strikes through it with its talons, leaving an untidy hole that may be surrounded by wing and tail-feather imprints. If it registers well, this 'strike' can be quite a sight. The feather imprints are made as the owl struggles to take off with possibly heavy prey. An ungraceful walker, it prefers to fly away from the scene.

You may stumble across a strike and guess that the owl's target could have been a vole (p. 88) scurrying around underneath the snow. Or you may be following the surface trail of an animal to find that it abruptly ends with this strike mark, where the animal has been seized by an owl, or perhaps a hawk or raven.

Similar Species: Other large, predatory birds also leave strike marks. If the prey left no approaching trail, the strike mark is likely an owl's, since owls hunt by sound. If there is a trail, the strike mark, usually with much sharper feather imprints, could be by a hawk or a Common Raven (p. 114), both of whom hunt by sight.

Common Raven

Print
Length: to 4 in (10 cm)

Straddle
to 4 in (10 cm)

Stride
Walking: to 6 in (15 cm)

Size
2 ft (60 cm)

COMMON RAVEN
Corvus corax

    This legendary bird spends a lot of time strutting around on the ground—confident behavior that may hint at its intelligence.

    Ravens show a typical alternating track pattern. Prints show three thick toes pointing forward and one toe pointing backward. When a Raven is in need of greater speed, perhaps for takeoff, it leaves a trail of diagonally placed pairs of prints that are rather irregular.

Similar Species: Other corvids also spend a lot of time poking around on the ground. Their tracks are similar, but smaller, and their strides are correspondingly shorter. For example, Black-billed Magpie (*Pica pica*) prints are up to 2 inches (5 cm) long.

Dark-eyed Junco

Print
Length: to 1.5 in (3.8 cm)
Straddle
1–1.5 in (2.5–3.8 cm)
Stride
Hopping: 1.5–5 in (3.8–13 cm)
Size
5.5–6.5 in (14–17 cm)

DARK-EYED JUNCO

Junco hyemalis

This common small bird typifies the many small hopping birds found in the state. Each foot has three forward-pointing toes and one longer toe at the rear. The best prints are left in snow, although in deep snow the toe detail is lost; the feet may show some dragging between the hops.

A good place to study this type of prints is near a birdfeeder. Watch the birds scurry around as they pick up fallen seeds, then have a look at the prints left behind. For example, juncos are attracted to seeds that chickadees scatter as they forage for sunflower seeds in the birdfeeder. Also look for tracks under coniferous trees, where juncos feed on fallen seeds in winter.

Similar Species: Toe size may help with identification—larger birds make larger prints—as can the season. In powdery snow, junco tracks could be mistaken for mouse (pp. 90–93) tracks; follow the trail to see if it disappears down a hole or into thin air.

Toads

Straddle
to 2.5 in (6.5 cm)

TOADS

Western Toad

The best place to look for toad tracks is undoubtedly along the muddy fringes of waterbodies, but they can occasionally be found in drier areas, for example as unclear trails in dusty patches of soil. In general, toads walk and frogs hop, but toads are pretty capable hoppers too, especially when being hassled by overly enthusiastic naturalists.

The only toad likely to be found in the state is the Western (Boreal) Toad (*Bufo boreas*), which inhabits streams, meadows and woodlands in southern Alaska. Toads leave rather abstract prints as they walk. The heels of the hind feet do not register. On less firm surfaces the toes often leave draglines.

Frogs

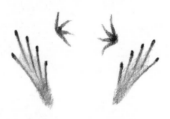

Straddle
to 3 in (7.5 cm)

FROGS

Wood Frog

As with toads, the best place to look for frog tracks is along the muddy fringes of waterbodies. The hardy Spotted Frog (*Rana pretiosa*) is found in southern Alaska, where it frequents cold mountain streams and lakes. The most widespread frog in Alaska, common even at high elevations, is the Wood Frog (*R. sylvatica*).

A frog's hopping action results in the two small forefeet registering in front of the long-toed hind prints. Although toads generally walk, toads are also capable hoppers. Frog tracks vary greatly in size, depending on species and age.

Salamanders and Newts

Straddle
to 7.5 cm (3 in)

SALAMANDERS AND NEWTS

Long-Toed Salamander

The harsh winters of Alaska are ill-suited to most amphibians, but a few hardy species of salamanders may be found in the southeastern portion of the state. With luck, the tracks of the Long-toed Salamander (*Ambystoma macrodactylum*) can be found near fresh water on the mainland and the coastal islands. Look for the tracks of this pretty salamander in soft mud under logs or under leafy debris at the edge of pools. Similar in size is the robust Northwestern Salamander (*A. gracile*), which can be found near lakes, ponds and slow-moving streams.

A slightly smaller close relative of the salamanders, the Rough-skinned Newt (*Taricha granulosa*), is found in the same habitats. If you happen to uncover one, it will strike a defensive posture intended to scare you away.

TRACK PATTERNS & PRINTS

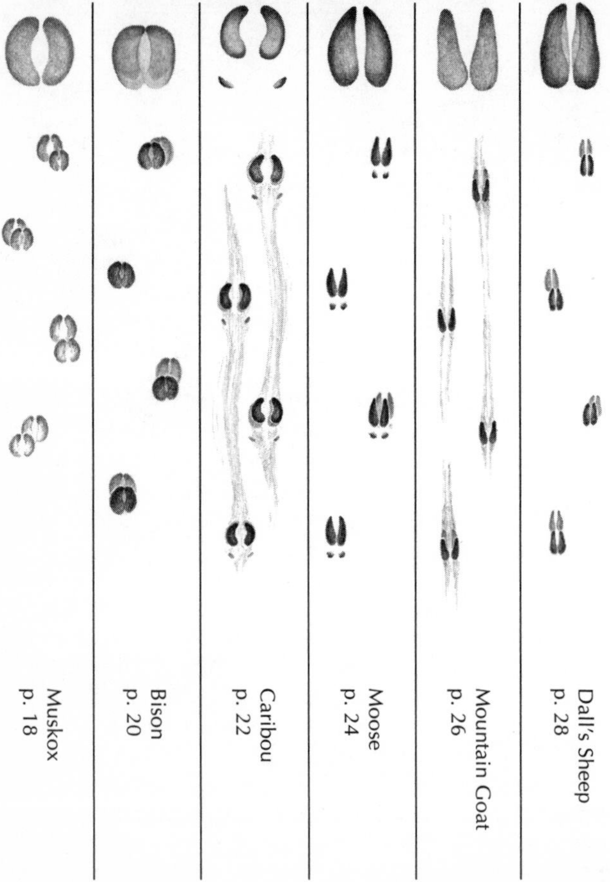

Muskox
p. 18

Bison
p. 20

Caribou
p. 22

Moose
p. 24

Mountain Goat
p. 26

Dall's Sheep
p. 28

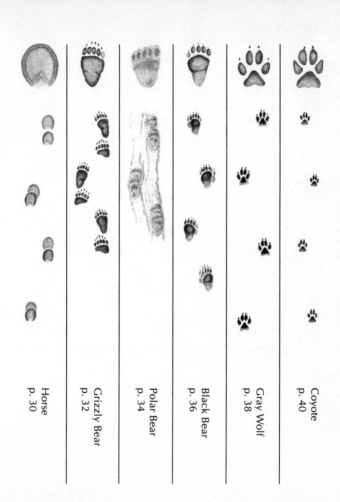

Horse
p. 30

Grizzly Bear
p. 32

Polar Bear
p. 34

Black Bear
p. 36

Gray Wolf
p. 38

Coyote
p. 40

TRACK PATTERNS & PRINTS

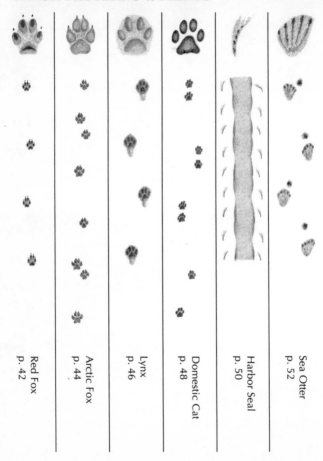

Red Fox
p. 42

Arctic Fox
p. 44

Lynx
p. 46

Domestic Cat
p. 48

Harbor Seal
p. 50

Sea Otter
p. 52

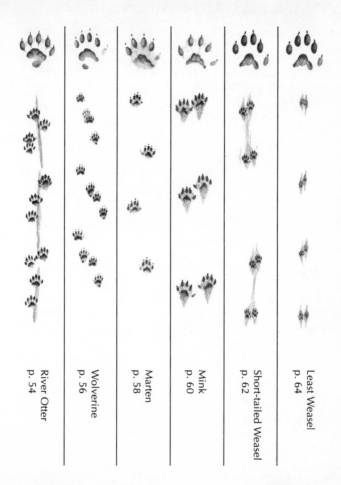

River Otter
p. 54

Wolverine
p. 56

Marten
p. 58

Mink
p. 60

Short-tailed Weasel
p. 62

Least Weasel
p. 64

TRACK PATTERNS & PRINTS

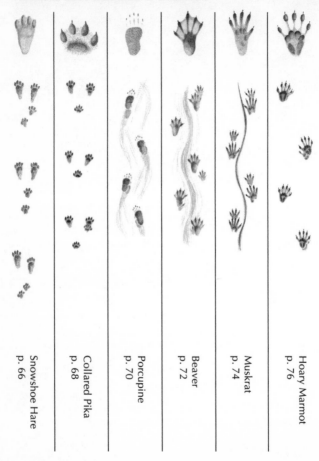

Snowshoe Hare
p. 66

Collared Pika
p. 68

Porcupine
p. 70

Beaver
p. 72

Muskrat
p. 74

Hoary Marmot
p. 76

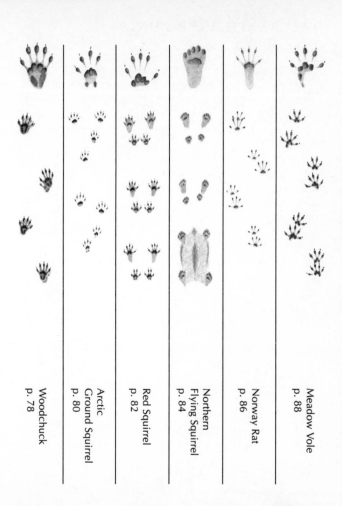

Woodchuck
p. 78

Arctic
Ground Squirrel
p. 80

Red Squirrel
p. 82

Northern
Flying Squirrel
p. 84

Norway Rat
p. 86

Meadow Vole
p. 88

TRACK PATTERNS & PRINTS

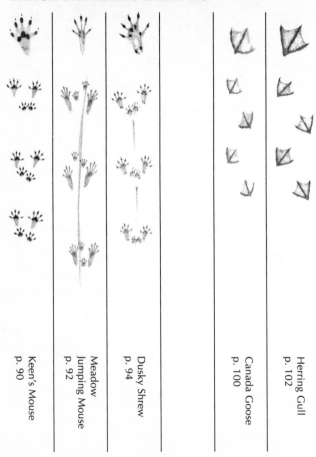

Keen's Mouse
p. 90

Meadow
Jumping Mouse
p. 92

Dusky Shrew
p. 94

Canada Goose
p. 100

Herring Gull
p. 102

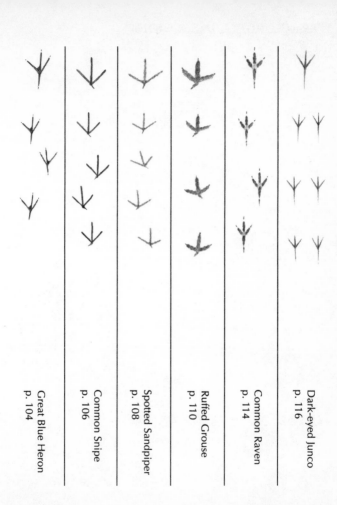

Great Blue Heron
p. 104

Common Snipe
p. 106

Spotted Sandpiper
p. 108

Ruffed Grouse
p. 110

Common Raven
p. 114

Dark-eyed Junco
p. 116

TRACK PATTERNS & PRINTS

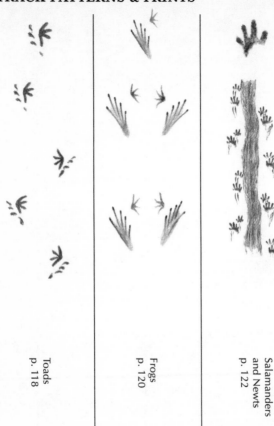

Toads
p. 118

Frogs
p. 120

Salamanders
and Newts
p. 122

HOOFED PRINTS

Caribou

Moose

Muskox

Mountain
Goat

Dall's
Sheep

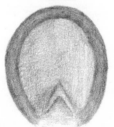

Horse

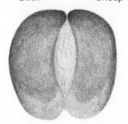

Bison

inch cm
0 — 0

1 —

2 — 5

FORE PRINTS

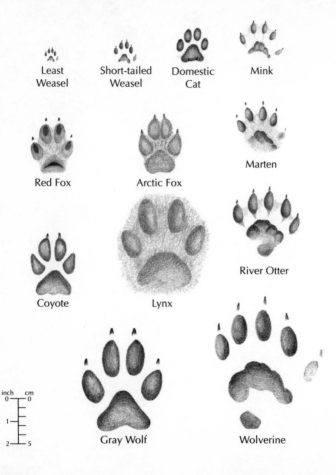

Least Weasel

Short-tailed Weasel

Domestic Cat

Mink

Red Fox

Arctic Fox

Marten

Coyote

Lynx

River Otter

Gray Wolf

Wolverine

inch cm
0   0

1

2   5

HIND PRINTS

Keen's Mouse

Meadow Vole

Dusky Shrew

inch   cm
0 —— 0

—— 1

—— 2

1 —
—— 3

—— 4

2 —— 5

Collared
Pika

Meadow
Jumping Mouse

Arctic Ground
Squirrel

HIND PRINTS

Norway
Rat

Northern
Flying
Squirrel

Red
Squirrel

Woodchuck

Hoary
Marmot

Muskrat

Porcupine

Snowshoe
Hare

inch   cm
0 —— 0

1 —
—— 5
2 ——

135

HIND PRINTS

Sea Otter

Beaver

Sasquatch

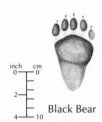

Black Bear

Grizzly Bear

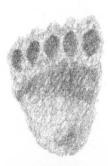

Polar Bear

inch | cm
0 — 0
2 —
4 — 10

BIBLIOGRAPHY

Behler, J.L., and F.W. King. 1979. *Field Guide to North American Reptiles and Amphibians*. National Audubon Society. New York: Alfred A. Knopf.

Burt, W.H. 1976. *A Field Guide to the Mammals*. Boston: Houghton Mifflin Company.

Farrand, J., Jr. 1995. *Familiar Animal Tracks of North America*. National Audubon Society Pocket Guide. New York: Alfred A. Knopf.

Forrest, L.R. 1988. *Field Guide to Tracking Animals in Snow*. Harrisburg: Stackpole Books.

Halfpenny, J. 1986. *A Field Guide to Mammal Tracking in North America*. Boulder: Johnson Publishing Company.

Headstrom, R. 1971. *Identifying Animal Tracks*. Toronto: General Publishing Company.

Murie, O.J. 1974. *A Field Guide to Animal Tracks*. The Peterson Field Guide Series. Boston: Houghton Mifflin Company.

Rezendes, P. 1992. *Tracking and the Art of Seeing: How to Read Animal Tracks and Signs*. Vermont: Camden House Publishing.

Scotter, G.W., and T.J. Ulrich. 1995. *Mammals of the Canadian Rockies*. Saskatoon: Fifth House.

Stall, C. 1989. *Animal Tracks of the Rocky Mountains*. Seattle: The Mountaineers.

Stokes, D., and L. Stokes. 1986. *A Guide to Animal Tracking and Behaviour*. Toronto: Little, Brown and Company.

Wassink, J.L. 1993. *Mammals of the Central Rockies*. Missoula: Mountain Press Publishing Company.

Whitaker, J.O., Jr. 1996. *National Audubon Society Field Guide to North American Mammals*. New York: Alfred A. Knopf.

INDEX

Page numbers in **bold typeface** refer to the primary (illustrated) treatments of animal species and their tracks.

ABOUT THE AUTHORS

Ian Sheldon, an accomplished artist, naturalist and educator, has lived in South Africa, Singapore, Britain and Canada. Caught collecting caterpillars at the age of three, he has been exposed to the beauty and diversity of nature ever since. He was educated at Cambridge University and the University of Alberta. When he is not in the tropics working on conservation projects or immersing himself in our beautiful wilderness, he is sharing his love for nature. Ian enjoys communicating this passion through the visual arts and the written word.

Tamara Hartson, equipped from the age of six with a canoe, a dip net and a note pad, grew up with a fascination for nature and the diversity of life. She has a degree in environmental conservation sciences and has photographed and written about biodiversity in Bermuda, the Galapagos Islands, the Amazon Basin, China, Tibet, Vietnam, Thailand and Malaysia.

LONE
PINE

Look for these other Lone Pine titles:

Backroads of Alaska and the Yukon
by Joan Donaldson-Yarmey
ISBN 1-55105-217-2 • 224 pages • B/W photographs • $11.95 U.S. •
$ 14.95 CDN

Explore North America's Final Frontier!
Starting at "mile 0" of the world famous Alaska highway, you can complete
any trip described in the book, in a day, or combine routes to form great
back-country tours. Several maps and photographs are included to help you
find your way.

Plants of Northern British Columbia–Revised & Updated!
Edited by Andy MacKinnon, Jim Pojar and Ray Coupé
ISBN 1-55105-108-7 • 352 pages • 570 color photographs, 600 illustrations •
$19.95 US • $26.95 CDN

More than 500 species of trees, shrubs, wildflowers, grasses, ferns, mosses
and lichens are illustrated and described. This comprehensive and
accessible guide to a region rich in plant life includes new photographs and
native uses of plants.

Seashore of British Columbia
by Ian Sheldon
ISBN 1-55105-163-X • 192 pages • 150 color illustrations • $11.95 US •
$15.95 CDN

A spectacular guide to British Columbia's intertidal life with full-color
illustrations of 150 species. Concise descriptions of mammals, cnidarians,
arthropods, worms, seaweeds and more.

Canadian Orders	**US Orders**
1-800-661-9017 Phone	1-800-518-3541 Phone
1-800-424-7173 Fax	1-800-548-1169 Fax

e-mail: info@lonepinepublishing.com